Instant Idea Book

Instant Science Lessons

for Elementary Teachers

- Biological Sciences
  (Pages 9-26)

- Earth Sciences
  (Pages 27-42)

- Physical Sciences
  (Pages 43-60)

(includes reproducible pages)

by

Barbara Gruber & Sue Gruber

Illustrations
Lynn Conklin Power

NOTICE! Reproduction of these materials for commercial resale, or for an entire school or school system is strictly prohibited. **Pages marked "reproducible" may be reproduced by the classroom teacher for classroom use only.** No other part of this publication may be reproduced for storage in a retrieval system, or transmitted, in any form or by any means—electronic, mechanical, recording, etc.—without the prior written permission of the publisher.

Copyright© 1986 Frank Schaffer Publications, Inc.
All rights reserved • Printed in the U.S.A.
Published by **Frank Schaffer Publications, Inc.**
1028 Via Mirabel, Palos Verdes Estates, California 90274

ISBN #0-86734-058-4

Table of Contents

INTRODUCTION 5
    Be a Scientific Observer 6
    Establish a Science Center 8

BIOLOGICAL SCIENCES 9
    The World of Plants 10-13
    The World of Animals 14-17
    Learning About the Human Body 18-22
    Ecosystems 23-26

EARTH SCIENCES 27
    Oceanography 28-31
    Geology 32-36
    Meteorology 37-39
    Astronomy 40-42

PHYSICAL SCIENCES 43
    Sound 44-47
    Gravity 48-50
    Light 51-53
    Magnetism 54-56
    Matter 57-60

REPRODUCIBLE PAGES
    My Scientific Observation 7
    Moon-Phases Flip Book 41
    Science Experiment 61
    Film Review 62
    Science Reading Report 62
    Science Awards 63

Introduction

Instant Science Lessons is designed to enrich your science program. It encourages students to **think** and **do** besides listen and watch. The activities and demonstrations are easy to implement in your classroom and enjoyable for students!

Useful reproducible pages are included that can be used with any science experiment, science film or filmstrip, and science reading assignment. These multi-purpose forms coordinate with any science topic.

You'll be delighted to discover heightened interest and enthusiasm for science in your classroom.

Barbara Gruber
Barbara Gruber

Sue Gruber
Sue Gruber

Be a Scientific Observer

What Is Scientific Observation?

Scientific observation includes using the five senses to describe something in an objective manner. Teach your students observation skills by using rocks, shells, leaves, twigs or weather as the subject. *(You can use the Scientific Observation form on page 7, or elicit information from your class and write it on the chalkboard.)*

Observing a Peanut

Materials: a peanut in its shell for each student

1. **See:**

    Have students describe the appearance of the peanuts: color
    size
    shape

    Then have students crack the peanuts and describe what they see.

2. **Touch:**

    Have students describe the way the peanuts feel: texture (rough or smooth)
    weight (light or heavy)

3. **Hear:**

    Have students describe the sound of the peanuts: when sitting on desk
    when shaken

4. **Smell:**

    Have students describe the smell of the peanuts:

    (Note: If students say the peanut smells "good," that is an opinion, not an observation.)

5. **Taste:**

    Have students taste the peanuts and describe the flavor: sweet
    sour
    bitter
    salty

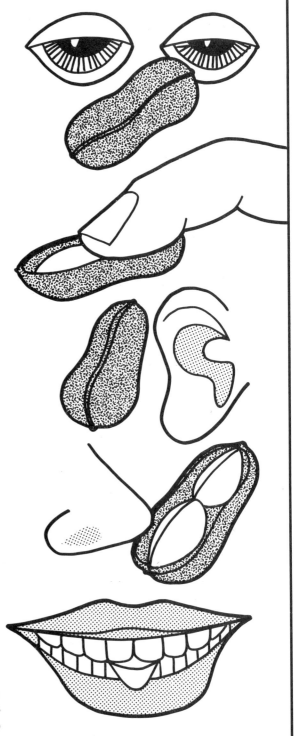

Note: Once students understand the observation process, give them an opportunity to independently observe objects and list their observations on paper before a group discussion.

Name _____ Date _____

My Scientific Observation

I am observing a(n) _____

1. **See:** _____
   _____
   _____

2. **Touch:** _____
   _____
   _____

3. **Hear:** _____
   _____
   _____

4. **Smell:** _____
   _____
   _____

5. **Taste:** _____
   _____
   _____

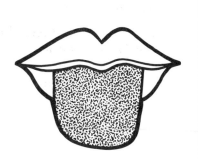

a reproducible page

© Frank Schaffer Publications, Inc.   FS-8308 Instant Idea Book

Establish a Science Center

Establish an area (table, desk, countertop) for a science center. Choose a different topic for every few weeks or each month.

Encourage students to bring books, pictures, collections, and specimens for the science center. Display books, magazines, objects, and posters at the center also. Have a suggestion box so students can suggest topics for the science center.

Biological Sciences

The Study of Living Things

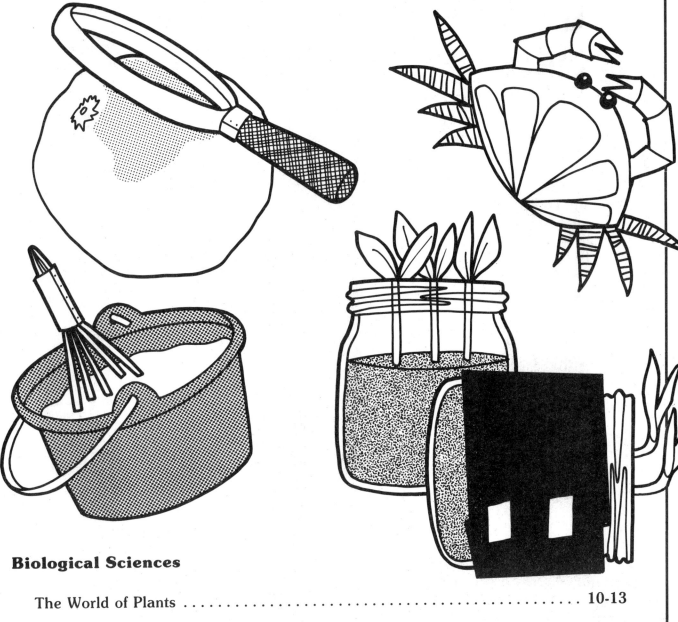

Biological Sciences

The World of Plants 10-13

The World of Animals 14-17

Learning About the Human Body 18-22

Ecosystems 23-26

The World of Plants

Where Do Seeds Come From?

- Objective: Students discover the natural source of seeds.
- Science concept: Each species can only produce offspring of the same species. For example, seeds from watermelons will only produce watermelons.
- Materials needed: packet of marigold seeds
  flowers from marigold plants
  potting soil
  small planting containers
  (cups, egg cartons, eggshell halves)

Ask your students where seeds come from. Jot a list of responses on the chalkboard.

Gently pull open the base of a marigold flower to reveal the cluster of seeds. Spread the seeds out on a paper towel, so they are visible to the students.

Compare the marigold seeds taken from the flower to the seeds in the packet.

Arrange your students in groups or partners, and let students gather seeds from additional marigold flowers.

Let students plant the seeds to watch how plants grow. Students will be excited to take their plants home and show their families where seeds really come from.

Search for Seeds

Ask students to bring samples of other kinds of seeds and the parent plants or fruits.

For example: slice of cucumber
orange slices
cherry
plum
squash

Label and display the seeds and plants or fruits.

The World of Plants

How Plants Help People

- Objective: Students recognize how plants are used in daily life.
- Science concept: Plants supply people with an important need. Plants add beauty to our lives.
- Materials needed:

    12" × 18" paper for each student
    scissors, paste, crayons
    magazines to cut apart

Ask students to look around the classroom for items that came from plants. Jot a list on the chalkboard of objects mentioned by students (wood pencils, desks, cotton clothing).

Ask students to think about the food they ate at lunchtime. Jot a list on the chalkboard of foods that came from plants (bread, fruit, lettuce, peanuts).

Have each student fold a paper into fourths and label the sections: Food, Clothing, Shelter, Other Uses. Students cut out pictures (or draw and label items) that show ways people use plants and paste the pictures in the appropriate section of the paper.

To make this a group activity, divide a bulletin board into four sections instead of having students work on individual papers.

- Research Questions:

How do plants give off oxygen?
How are plants used in medicines?
How did plants help form coal, oil and natural gas?
How were plants used by Native Americans?

Time Lapse Illustration

Plant seeds in jars or in a terrarium so the seeds can be seen underground through the glass. Have students observe one seed over a period of time and draw the changes they observed.

The World of Plants

Growing Your Own Mold

Molds need air, food and water. They grow best where it is warm and not too light. Get an orange, lemon or grapefruit that has spots of greenish mold. Have students look at the green fuzzy spot on the orange with a magnifying glass.

Have students grow their own mold by putting slices of bread separately inside plastic bags. Students brush some of the green dust from the moldy orange into their bags. Then leaving some air trapped inside, students tightly close their bags. The dust they brush on the bread is spores (like seeds) which will cause that kind of mold to grow on the bread.

Have each student record her experiment.

I put the spores on the bread. _____   date   I saw the first spot of mold. _____   date   The mold is growing all over the bread. _____   date   I examined the mold on the bread with a hand lens and saw _____.

You can grow mold on bread under different conditions to learn in which environment it grows best. Place some specimens in bright sunlight, in darkness, and in a refrigerator.

Heading for the Sky

Demonstrate to your class that seeds always grow upward. Put soil into two jars. Plant a few bean seeds in each jar. Tape black construction paper around each jar. Place jars on a sunny windowsill leaving one jar upright and the other on its side. Students will observe that the seedlings all grow toward the sunlight.

To make seeds sprout faster, soak them in water overnight before planting.

©Frank Schaffer Publications, Inc.

FS-8308 Instant Idea Book

The World of Plants

How to Collect Plants

Make leaf prints by pressing each leaf on a stamp pad. Place a piece of paper over the leaf to keep your hands clean. Be certain to press firmly on all parts of the leaf. Pick the leaf up with tweezers and place it on the paper in the location where you want the print. Use scrap paper to press the inky leaf on the paper. Pick the leaf up with tweezers and discard. Label each print.

Make leaf prints using a photocopy machine. When copying light-colored leaves, place the leaves on the copy surface, and then cover the leaves with a piece of black paper, so the leaves stand out on the copy. This will make the background dark, instead of white.

Press flowers and plants between newspaper or paper towels and cover with a stack of heavy books. You can arrange the specimen on a piece of paper and affix it with a few spots of white glue before pressing. After pressing your specimen, it will already be mounted.

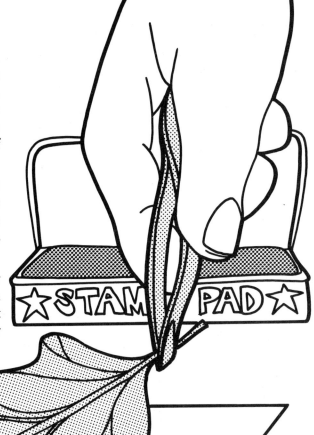

Handy-Dandy Bags

Plastic bags come in handy for growing seeds. Put a handful of very moist soil in a bag and add seeds. Seal the bag, trapping some air inside. Then place it in a dark place until the seeds sprout. Next, open the top of the bag and place it in a sunny spot to watch your plants grow.

The World of Animals

Animal Posters

Post six large sheets of butcher paper or tagboard (approximately 36" x 40"). Write a different category of animals on each poster. Have students paste pictures of animals on the correct poster. Each animal picture should be labelled.

Insects	Reptiles
Fish	Birds
Amphibians	Mammals
Extinct & Prehistoric Animals	

Primary students can sort animals into these categories:

Farm Animals	Zoo Animals
Pets	Wild Animals
Tame Animals	

Intermediate students can add an endangered animals category.

Animal Identification Contest

Post pictures of a variety of animals on a bulletin board. Number each animal that is posted. Tell students that they are to number their papers and to identify as many animals as possible.

Name _____
1. kangaroo   7. otter
2. roadrunner 8. reindeer
3. ape   9. koala
4. monkey   10. lynx
5. camel   11. fox
6. seal   12. hare

Vocabulary Words for Learning About Animals

Discuss the meaning of these words. Have students take a spelling test or write sentences using these words.

migrate—(seasonal or periodic movement to another region)

hibernate—(to be in an inactive or dormant state)

habitat—(the type of environment where a living thing is normally found)

extinct—(no longer living)

prehistoric—(from an era before recorded history)

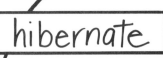

© Frank Schaffer Publications, Inc.   FS-8308 Instant Idea Book

The World of Animals

Observation Box

Convert an aquarium into a terrarium. Put a layer of charcoal and pebbles in the bottom of the tank. Add a layer of soil and plant some small plants. Add a few rocks and twigs. Cover the tank with a lid or screen. Place insects or spiders in the terrarium. After observing them for a short while, release them to their natural environment.

Animal Defenses

Animals protect themselves in a variety of ways. Have students make posters showing how animals protect themselves.

rattlesnake—fangs
boa constrictor—muscular body to squeeze prey
cat—claws
skunk—horrible-smelling liquid to spray at prey
dog—teeth
crab—claws
porcupine—quills
bear—claws, teeth, jaws
deer—sharp hind hooves and antlers to harm prey
octopus—tentacles
lion—claws, teeth, jaws
baboon—sharp teeth to bite prey
hawk—strong claws and curved beak
opossum—plays dead
clams—hard shell
kangaroo—claw on one toe to kick prey
wild pigs—sharp tusks
bees—stinger

Have students work independently or with a partner. Assign each pair of students one animal to illustrate. Post the illustrations on a bulletin board or staple into a booklet called "How Animals Protect Themselves."

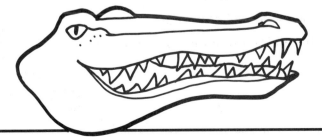

© Frank Schaffer Publications, Inc.   FS-8308 Instant Idea Book

The World of Animals

Frog for a Day

Have each student choose a different animal to write about. Students write about a typical twenty-four-hour period in the lives of their animals.

Students should include:

- where the animal lives
- what it eats
- how it protects itself
- how it travels (land, air, water)

Ancient Footprints

Make plaster casts of impressions in clay to show students how fossils are formed. Cover the bottom of a small box or plastic dish with modeling clay (small margarine tubs work fine). Press a small seashell firmly into the clay, so a clear imprint remains when the shell is removed. Then add water to plaster of Paris until it is a creamy consistency. Pour the plaster of Paris mixture over the imprint to fill the shell print. After the plaster has hardened, remove it and you will have a cast of the original shell.

This technique is used by paleontologists to make molds of fossil animal shells and footprints.

Long, Long Ago

Children are fascinated with prehistoric animals. Gather books and pictures of prehistoric animals. Let each student use clay to make a model of a prehistoric animal. Each model should be labelled.

A diorama scene can be made by placing clay animals in a shoebox.

The World of Animals

Insects and Spiders

Insects have six legs and three body parts.
Spiders have eight legs and two body parts.

Have students draw and label the parts of an insect and a spider.

Life Cycles

Have students make a life-cycle chart showing the life cycle of an animal going through its life stages.

For example:

- silkworm

    moth lays eggs
    larvae (silkworms)
    adult silkworm
    pupa (cocoon)
    moth

- frog

    egg
    tadpole
    adult frog

- moth

    egg
    larva (caterpillar)
    pupa (chrysalis)
    adult moth

Learning About the Human Body

Our Sensational Senses

The five senses help people to be aware of everything in the world.

Sight

Eyes have receptors for color. This activity will help your students understand how the eye perceives color.

Materials needed: crayons
white drawing paper

Directions for students:

1. Draw a square that is approximately four inches in size in the center of a piece of white paper. Color the square heavily with a red crayon.

2. Stare intently at the red square for one to two minutes. Then shift your eyes to a plain white piece of paper. You will see a green square on the plain white paper. What you are seeing is called an afterimage.

The reason students see a green square is that the color receptors for red are tired from staring at the red square. When students look at the plain white paper, light coming from the paper contains all the colors, but the red receptors are too strained to detect red. When the eye becomes insensitive to a certain color, it sees that color's complement. Try this with other colors, also.

Touch

The skin feels many different sensations which tell about objects.

Materials needed: bag or pillowcase
objects of varying textures to touch

1. Let students take turns putting their hands in the bag. Have each person pick up one object inside the bag and describe the way it feels.

2. To vary the activity, have students remove certain categories of objects.

   For example: smooth objects
   rough objects
   cold objects
   warm objects
   soft objects
   hard objects

©Frank Schaffer Publications, Inc.　　　18　　　FS-8308 Instant Idea Book

Learning About the Human Body

Taste

The sense of taste and smell work together. However, the tongue has specific areas that detect different tastes.

Materials needed: crayons
pencil
8 1/2" x 11" white paper

As the students are making a tongue map on their papers, you should be making one on the chalkboard.

Directions for students:

1. Draw a tongue in the center of your paper. Color the tongue pink. Label the back and tip of the tongue.

2. Use symbols to indicate the different taste regions on the tongue as shown. Be sure to include a key to interpret the symbols.

3. Test these taste areas for yourselves. Put a small amount of salt on the side of your tongue toward the back. It will taste sour. Then put a small amount of salt on the side again but this time move closer to the front of your tongue. It should taste salty. Do this using sugar, lemon juice and orange peel.

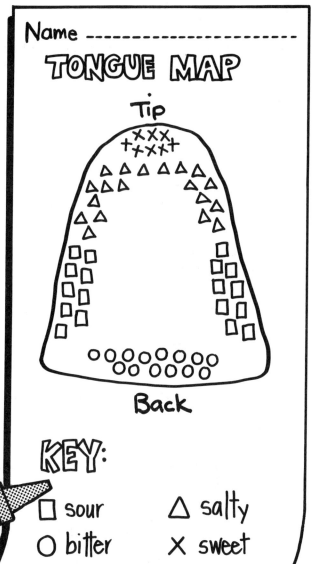

Hearing

The world is filled with sounds. Even when you think it is quiet, there are background noises.

Have students sit as quietly as possible for two minutes. Tell them to listen carefully for sounds. When the time is up, have students write the sounds they heard.

Smell

There are sensors inside your nose that detect different smells in the air.

Materials needed: various spices and flavoring extracts
a blindfold

Have students take turns identifying spices and extracts by their smell.

Learning About the Human Body

Your Heart Is a Pump

The heart beats at different rates depending on a person's level of activity. Students can measure their heartbeats at rest and after an activity.

Materials needed: watch or clock with second hand
　　　　　　　　　　paper and pencil

1. Each student prepares a record sheet.

2. Show students how to find their pulse points in the neck and wrist. Give students a signal to begin counting their heartbeats. They count their heartbeats for 10 seconds and record their resting heartbeats. Students can multiply this count by six to get the number of heartbeats per minute.

3. Have students jump in place for approximately a minute. Immediately following jumping, have students measure and record their active heartbeats.

My Record Sheet

Name _____

My Pulse

Resting _____

Active _____

Skin—Your Largest Organ

Skin is the largest organ in the human body. It is the most important factor in helping the body regulate its temperature. The body gives off moisture through the skin. To demonstrate this process have students perform the following experiment.

Materials needed: plastic bag for each student's hand
　　　　　　　　　　string or rubber bands

1. Have students place one hand in the bag and secure it gently around the wrist with the rubber band.

2. Have students do some form of excercise— jumping jacks, toe touches or leg lifts.

3. Following the exercise, have students look at the bags on their hands and describe what they see and feel. (The inside of the bags will be moist and warm.)

Learning About the Human Body

Right or Left?

Most people know if they are right- or left-handed. Each person also has an eye and foot preference. This preference is determined by the brain. Students can discover their preferences with this lesson.

Materials needed: paper and pencil
a tissue or paper towel for each student

1. Have each student prepare his worksheet as shown.

2. Hand Preference: On their worksheets have students write *right* or *left* to tell which hand they use for writing.

3. Eye Preference: Students use pencils to poke one hole in the center of their tissues. Tell students to pick up their tissues and look through the hole with one eye. The eye that they held the tissues up to is the eye their brain prefers to use. In other words, they automatically held it up to that eye. They record right or left eye preference on the worksheet.

4. Foot Preference: Have students crumple their tissues into a ball, stand up and drop them on the floor. Tell students to gently kick the tissues. The foot used is the preferred foot. They record right or left on the worksheet.

Results can be tabulated on a graph!

Name _____
Date _____

Right or Left?

Hand _____
Eye _____
Foot _____

Eye, Foot, and Hand Preference Graph for Our Class

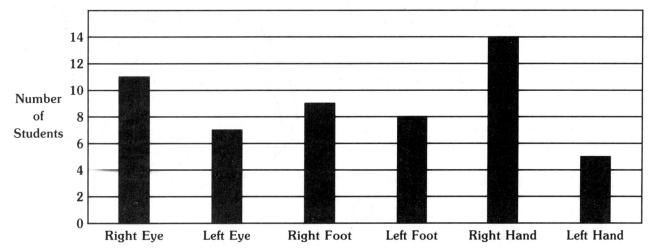

Now that students know how to determine the brain's preferred eye and foot, they can test their families and friends, too!

Learning About the Human Body

Food—Your Body's Fuel

Use actual food products your students consume to establish good eating habits.

Materials needed: empty containers/boxes/cans from a variety of foods
counter or tabletop area

1. Have students bring clean, empty food containers from home.

2. When you have gathered ten or more containers, divide a countertop into five sections. Label the sections: Meats and Proteins, Fruits and Vegetables, Breads and Cereals, Dairy Products, Other Foods.

3. Review the kinds of food that belong in each of the food groups. Tell students that certain snack foods like candy and soft drinks don't fit into the traditional four categories. They belong in the "other foods" group.

4. Sort the food containers into the five categories. Sorting can be done as a whole class, small group or individual activity.

Ecosystems

Food Webs

To demonstrate the interdependency of living things, have your students participate in food-web simulation.

Materials needed: 8 1/2" × 11" paper for each student

1. Explain to the class that living things depend on one another for survival. Ask students to think about a natural setting where plants and animals live together such as a forest.

2. Elicit from the class plants and animals that would be found in this environment. List plants and animals on the chalkboard. For example: fish, mice, squirrels, rabbits, owls, hawks, foxes, raccoons, grass, trees . . . Be sure to add sun, water and air which are needed to sustain living things.

3. Give each student a paper. Ask one student to represent the sun. Have that student write *sun* on her paper and stand in the front of the room holding her paper sign. Assign each of the other students one component of the environment. Students stand in these groups:
    plant-eating animals
    meat-eating animals
    air
    water
    plants and trees

4. Introduce a problem into "the environment."

    Problem: polluted water
    This affects plants and trees.
        (Have students representing water, plants and trees sit down.)

    Ask: Who depends on plants and trees?
        (Plant-eating animals depend on them, so have those students be seated.)

    Continue this until all or most students are seated to illustrate interdependency of living things in a food-web.

Discuss other potential problems such as insecticides, air pollution, animal diseases and their effects on the community of living things.

5. Following this activity, have students draw a food web on the other side of their paper signs.

Ecosystems

How People Affect the Environment

Materials needed:

12" × 18" drawing paper for each student (or pair of students)

1. Each student draws a picture of a natural environment (seashore, forest, desert, valley). Tell students they have 15 minutes in which to draw these natural settings. (There will be additional time later for coloring the pictures).

2. After 15 minutes, ask students how human beings could negatively affect the environments they have drawn. (Littering, building too many roads or buildings, trampling plants, over-hunting wildlife, polluting the water or air....) List these ideas on the chalkboard.

3. On the backs of their pictures, students copy the effects listed that relate to their environments.

4. Then ask students to brainstorm ways to prevent these negative effects. (Handy trash cans, instructional road signs, buildings which take the environment into consideration....)

5. Then students draw those items which will help prevent pollution and destruction in their environments.

The Living Desert

Materials needed:
- jar/bottle/aquarium
- gravel
- sand
- potting soil
- small cactus plants

Put a layer of gravel in the bottom of the terrarium. Mix sand and potting soil making a mixture that is seventy-five percent sand. Spread a layer of this mixture that is approximately two inches in depth over the gravel. Plant cacti and add a few rocks and twigs. Give plants a small amount of water and cover the terrarium. If condensation appears, remove the lid for a few hours.

Ecosystems

Nature's Way of Recycling

This activity will demonstrate the importance of recycling materials that are not decomposed by nature.

Materials needed: one clay flowerpot
soil
litter

1. Cover the hole in the bottom of the flowerpot with a stone. Fill the pot approximately one-third full with soil.

2. Have students collect small samples of litter such as paper, styrofoam, foil and leaves. Also have students save a few bits of food scraps such as orange peel, bread, piece of apple from their lunches.

3. Add a layer of the litter to the flower pot and cover it with soil until the pot is almost full. Sprinkle it with water, so the soil is moist but not too wet. Cover with plastic wrap and place in a warm, dark place. Add water occasionally, so the soil stays moist.

4. After three weeks, dump out the contents of the flower pot on a thick layer of newspaper. Use a trowel to spread out the soil, so students can see which materials decomposed.

5. Discuss the importance of recycling trash. Tell the students where recycling centers are located in your community.

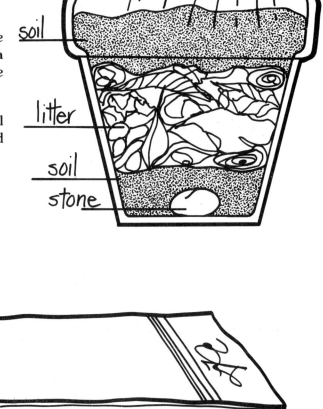

Note: "The World of Plants" pages 10 to 13, and "The World of Animals" pages 14 to 17 contain activities that relate to the study of ecosystems.

Ecosystems

Paper Making

Making paper is a wonderful way to teach your students how paper is recycled!

Materials needed: newspapers
bucket
cornstarch
whisk or non-electric egg beater
plastic wrap
piece of window screen
rolling pin or can

1. Fill bucket half-full with newspaper that is torn in small pieces. Add enough water to thoroughly wet the newspaper and let it soak for several hours.

2. Beat the paper/water mixture into a creamy pulp.

3. Dissolve three tablespoons of cornstarch in a cup of water and then add it to the mixture in the bucket. Stir thoroughly.

4. Lower a piece of window screen into the bucket of pulp repeatedly until it is covered with a layer of the mixture. This layer should be about 1/8" thick.

5. Place the screen on a thick layer of newspaper. Cover with plastic wrap and press out the extra moisture with a rolling pin (or use a can like a rolling pin).

6. Prop the pulp-covered screen in a place where it can dry. After the fibers dry (about one or two days), peel the recycled paper away from the screen.

Earth Sciences

The Study of What the Earth Is Made Of

EARTH SCIENCES

Oceanography 28-31

Geology (Study of the earth's crust) 32-36

Meteorology (Study of weather) 37-39

Astronomy (Study of sun, stars, planets) 40-42

Oceanography

Whale Models

Provide books about whales. Have each student make a model of a particular kind of whale using modeling clay. Make a display of whale models in your classroom. Students label their models with each whale's name and size.

Older students can make their models to scale (ten feet being equal to one inch in the length of the clay model). The model of the Blue Whale, which can grow to 100 feet, would be ten inches long, the Sperm Whale model would be six inches long because that whale grows to 60 feet long.

This activity could be done using sharks, dolphins or porpoises.

Deep-Sea Diver

Cut pictures from magazines of things that live in the ocean and post them on a bulletin board. Number each object. Tell students to pretend they are deep-sea divers who have spotted these living things under the sea. Students write deep-sea adventures describing any five ocean creatures in their stories.

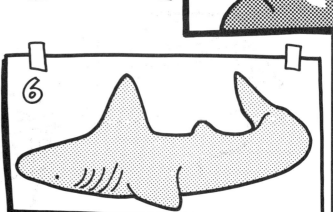

Oceanography

The Underwater World

Cover a bulletin board with blue paper. Cut a brown paper "ocean floor" and green paper "plants." Have each student draw, color, and cut out something that lives in the ocean and pin it to the underwater-mural bulletin board.

Make a Water Scope!

Show your students how to make a water scope so they can view the underwater world. Cut a circle from the bottom of a small, plastic bucket. Use a rubber band to cover the other end of the bucket with a piece of heavy, clear plastic wrap (or plastic from a plastic bag). Use the water scope to view underwater life in ponds, creeks and tidepools.

Oceanography

Water Pressure

When something goes into water, the water is pushed aside and the level of the water rises. Have a student put her hand inside a plastic bag. Put a rubber band gently around her wrist. Then dip her hand into a deep pan of water. She will feel the water pressing on her hand harder as she pushes deeper into the water.

Pressure In the Deep

In the ocean, water pressure increases with depth. There is more pressure at deeper levels because the weight of all the water above is pressing down on the water below. Demonstrate this by using a hammer and nail to poke three holes in a tall can. Set the can in or beside a sink and pour water into it. Water will spurt from all three holes, but it will spurt farthest from the hole at the bottom because the pressure is greater at that level.

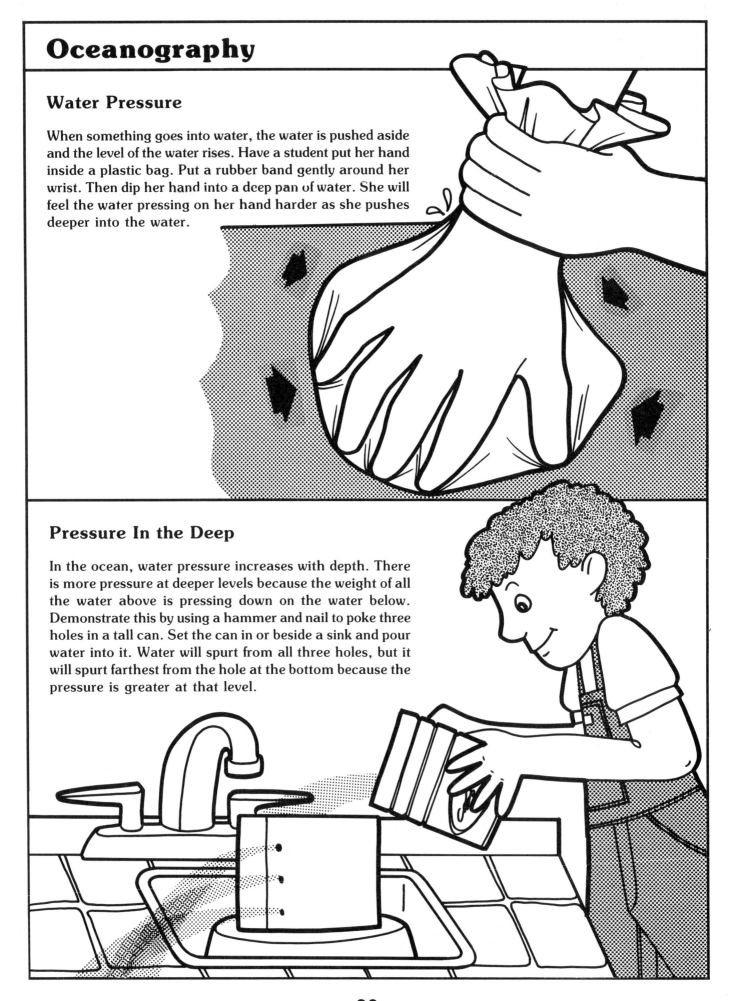

Oceanography

How Water Affects the Weather

Water stores heat better than the land does. Therefore, the air in winter is a few degrees warmer in areas near oceans and lakes.

Demonstrate that water stores heat more effectively than land by this easy experiment.

Materials needed:
- two metal cans the same size
- dirt
- water
- a meat thermometer

Fill one can with water and the other can to the same level with soil. Place both cans in direct sunlight for several hours (or in an oven set to warm for an hour). Remove both cans from the sun and place them where they can cool. Record the temperature of each can every fifteen minutes for the next hour.

TIME	SOIL	WATER
1:15	___	___
1:30	___	___
1:45	___	___
2:00	___	___
2:15	___	___
2:30	___	___

Sea Shells

Display sea shells and books about shells. Have your students make observations about a seashell. *(Use the reproducible scientific observation sheet on page 7.)*

Geology

Investigating Sedimentary Rocks

Sedimentary rocks are composed of different layers. These layers can be pebbles, sand, silt or fossil fragments. Your students will understand the characteristics of sedimentary rocks by making a model of sedimentary layers.

Materials needed: glass jars with lids
water
rocks
pebbles
sand
dirt

This activity can be done individually, in small groups or as a demonstration by the teacher for the class.

1. Students collect rocks, pebbles, sand and dirt at school or home.

2. Fill a jar one-third full with the rocks, pebbles, sand and dirt. Add water and shake the jar vigorously.

3. Let the jar stand undisturbed overnight. Materials will settle into layers with the coarser, heavier materials at the bottom and the finer, lighter layers at the top.

4. Tell the class that these sedimentary layers usually form in lake beds and near running water. After a long period of time, they turn to rock.

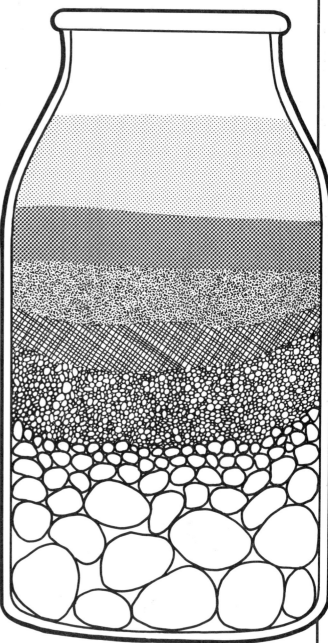

Rock Specimens

Encourage students to bring rocks to school for the science center. *(See page 8 for how to set up a science center. Students can do scientific observations of rocks using the process described on pages 6 and 7.)*

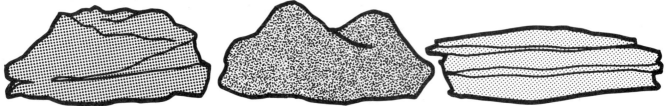

Geology

How Mountains Are Formed

Let students discover how mountains are formed by shifts in the earth's crust that force layers of rock to bulge.

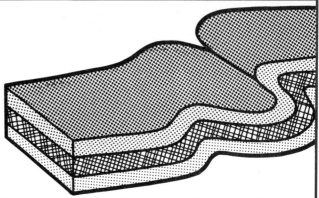

Materials needed: modeling clay in two or more colors for each child or group of children

1. Shape the clay into flat strips as shown. Layer alternating colors of strips so there are three or more layers. These strips represent layers of rocks.

2. Slowly push the opposite edges of the strips together causing them to bulge in the middle. This shows what happens to layers of rock when mountains are formed.

3. Flatten the "mountain" back into original strip arrangement and push one edge toward the center, instead of both edges. Now a different folding pattern of the layers will occur. In nature, the forces that result in mountain building usually put uneven pressure on the rock layers.

How Erosion Shapes the Land

The forces of erosion are strong. Through erosion Arizona's Grand Canyon was formed. This demonstration will show students how erosion occurs.

Materials needed: container of water
dirt
tagboard (approximately 12" × 18")
bucket/dishpan

1. Fold the tagboard and make a small "mountain" of dirt as shown.

2. Hold the tagboard at a slight incline on the edge of the dishpan. Slowly pour water on the top of the dirt mountain. The water will carry the dirt downstream into the dishpan. In nature, mountains are slowly eroded by water.

3. Discuss with your class how rain and running water constantly change the landscape.

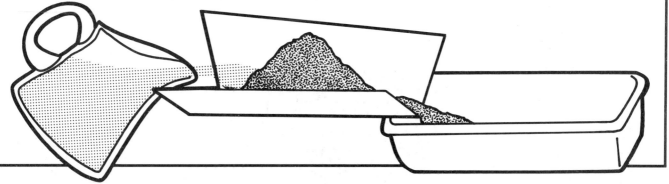

Geology

How a Glacier Moves

A glacier is a large mass of slowly moving ice. Glaciers can be so large that they cover an entire continent. Within the glacier, ice is melting and then refreezing. This melting-refreezing action causes the glacier to move slowly.

This experiment will show students how ice under pressure can melt and refreeze.

Materials needed: blocks of ice (freeze water in a one-quart milk carton)
thin wire or fishing line
two fishing weights at least six ounces each
two identical containers on which to rest the ice block
   (two coffee or juice cans)

1. Thread the fishing weights on the wire. Twist or tie the ends to form a loop as shown.

2. Place the loop of wire around the ice block with the weights hanging underneath.

3. Suspend the block of ice between the two containers with the weights hanging freely. The wire will move through the block of ice and eventually will fall out the bottom without breaking the block of ice. The weight of the wire on the ice block melts a path allowing the wire to pass. Then that path refreezes after the wire has passed.

Ice Is Powerful

Water seeps into small cracks in rocks. When the water freezes, it expands and makes the cracks larger. Eventually the rocks break apart from the freezing action of the ice. This will demonstrate to students how water expands when frozen.

Materials needed: plastic container with lid

1. Completely fill the container with water. Cover the container.

2. Place the container in a freezer overnight. Show the class the container of ice. The lid will be raised and the sides may be pushed out further.

3. Let the ice melt and the lid will fit back on the container.

Geology

Growing Crystals

All rocks are made of minerals. Minerals grow in certain shapes called crystals. Growing a string of crystals is an exciting project for the classroom.

Materials needed: sugar
very hot water
string
paper clip
pencil
glass jar
metal spoon

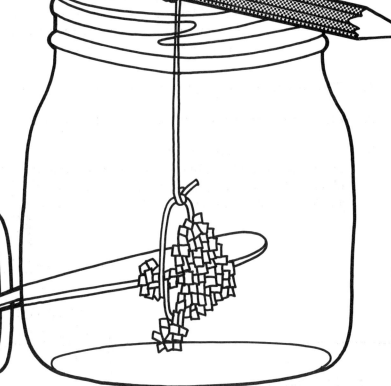

1. Pour hot water (heated on a stove, if possible) into a glass jar. To avoid having the jar break, place a metal spoon in the jar and slowly pour the hot water. Remove the spoon.

2. Add several spoonfuls of sugar and stir till the sugar dissolves. Continue adding sugar until no more sugar will dissolve in the hot water.

3. Tie a paper clip on one end of a string. Tie the other end of the string around the center of a pencil and place it across the top of the jar as shown. The paper clip is suspended in the sugar solution.

4. Set the jar in a place where it will not be disturbed, so crystals can grow on the string. Crystals will be visible within a few days. They will continue to grow as long as the solution contains dissolved sugar.

5. Remove the string of crystals from the jar. Have students look at the structure of the crystals with a hand lens. Compare these crystals grown in the jar to granulated sugar. (The crystal structure will be the same.)

You can use the same process to grow crystals from salt or Epsom salts.

Geology

Grow Stalagmites and Stalactites!

Stalagmites and stalactites are made of crystals. They occur in caves. Stalagmites grow up from the cave floor and stalactites grow down from the ceiling of the cave. Your students will enjoy watching stalagmites and stalactites grow in this experiment.

Materials needed: two jars
washing soda (buy at a grocery store in cleaning products section)
spoon
very hot water
coffee can lid (or old plate)
five 15" pieces of wool yarn twisted together to make one thick strand

1. Fill the jars with very hot water. Stir in washing soda until no more will dissolve in the water.

2. Place the jars about eight to ten inches apart in a warm location where they will not be disturbed. Place the lid or plate in between the jars.

3. Dip the yarn into the washing soda solution in one of the jars. Then drape the yarn between the two jars as shown. The solution will drip from the yarn onto the lid. A stalagmite will form on the lid and a stalactite will hang down from the yarn. These two will touch and form a column in a few days.

Meteorology

Evaporation

When water dries up, it turns into tiny droplets that go into the air. They are so small that you cannot see them. The water in the air is called water vapor. It rises into the sky and joins with other drops of water to form clouds. When water turns into water vapor, the change is called evaporation. Show students how water evaporates faster in the sun.

Put the same amount of water on two plates. Place one on a sunny windowsill and the other in a shady spot. You will see that the water disappears at a faster rate from the plate in the sun.

The Water Cycle

There are three steps in the water cycle:

1. Evaporation occurs when water turns to water vapor.

2. Condensation occurs when the water vapor changes back into tiny drops of water.

3. Precipitation takes place when the water drops fall to the earth again in the form of rain, hail or snow.

Demonstrate the water cycle for your students. On a sunny day, invert an empty glass on the grass. Water will evaporate from the grass into the glass. Then the water will condense into droplets. Drops of water will run down the sides of the glass and the water will go back to the earth.

Meterology

Types of Clouds

Clouds can help us predict the weather. Have your students make samples of different types of clouds using cotton. Students put small amounts of paste or glue on dark blue construction paper to fasten the cotton clouds. They brush pieces of charcoal across the stratus clouds to make them gray.

Cirrus clouds are feathery wisps of icy clouds high in the sky.

Cumulonimbus are cumulus clouds that are thick and heavy with water. A thunderhead builds up and looks like a puffy mountain on top of the cloud. Cumulonimbus clouds are a signal that a storm is coming.

Cumulus clouds are puffy white clouds.

Stratus clouds are spread out like thin sheets over a wide area of the sky at a low elevation. Light rain often falls from stratus clouds.

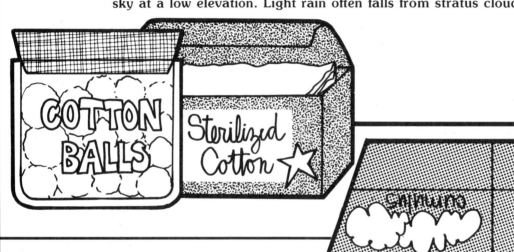

Weather and Seasons Affect Living Things

Have students divide a paper into four sections. Students label the sections for the seasons of the year. They each draw a picture to illustrate one day for each season.

Meteorology

The Changing Seasons

The changing seasons are caused by the position of the earth as it circles the sun. Different places on the earth receive different amounts of sunlight during the year. Help students understand this idea with this activity.

Materials needed: one yellow paper circle about 6" in diameter for each student (to represent the sun)
one light-blue paper circle about 4" in diameter for each student (to represent the earth)
(Make a "sun" and "earth" for yourself.)

1. Each student labels the yellow circle "sun."

2. Every student folds the blue circle in half, creases it and opens it up. On the fold line she draws a pencil line across the circle to represent the equator and writes the word "equator" on the line.

3. On the blue circle which represents the earth, each student labels north and south as shown below. She places an **X** to indicate the hemisphere in which she lives. (For example, if you live in the United States the **X** would appear in the northern hemisphere as shown.)

Demonstrate to students how the earth revolves around the sun by taping your "sun" to the chalkboard and moving the "earth" around it in a counterclockwise direction. Tell students it takes a year for the earth to journey around the sun. Show students that the earth is tilted in a position that does not change as it travels around the sun. After your demonstration, have students move their earths around their suns on their desktops.

Now, show students that when the earth is to the right of the sun, most of the northern hemisphere is tilted away from the sun's warming rays. Therefore, it is winter in the northern hemisphere and summer in the southern hemisphere. Have your students place their earths in this position. They continue to move their earths around their suns to show the position of the earth in different seasons of the year.

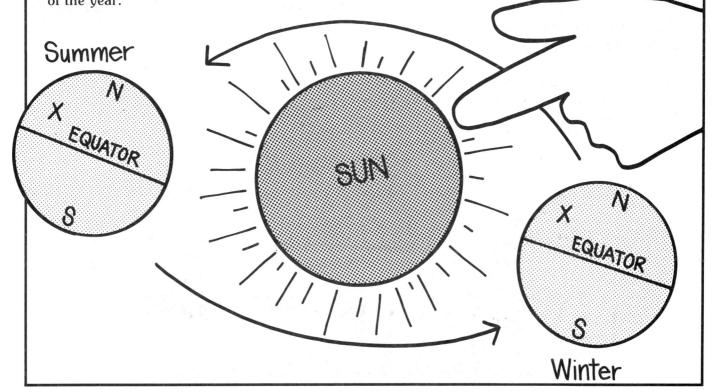

Astronomy

Sunny Day Shadows

Material needed: chalk

1. Early in the school day, take students outdoors to a sunny location on the playground. Bring a piece of chalk.

2. Stand where students can see you and your shadow. Ask a student to trace your shadow on the playground surface. Mark with an **X** where you were standing.

3. At noon, repeat this by standing in the same position on the playground and draw the shadow again with chalk.

4. Repeat again in the afternoon before students go home. On the playground, look at the three different shadow outlines. The path of the sun across the sky caused the difference in the direction and size of the shadows.

Moon-Phases Flip Book

Materials needed: a 14-page flip book for each student
    (Reproduce page 41 and cut each sheet into four flip book pages.)
    construction paper cover for flip book
    black crayons

1. Give each student a flip book containing 14 pages. Students write "Phases of the Moon" and their names on the front covers.

2. Tell your students that the moon's rotation around the earth takes 28 days. During its journey around the earth, it appears that the shape of the moon changes. However, the shape is not changing; you are seeing different portions of the moon.

3. At home, students along with their parents can observe the moon at approximately the same time in the evening every other night for 28 days. Each time a student observes the moon, she records the date and uses a black crayon to shade the circle as shown on each page of her flip book.

4. When the booklets are completed, students bring them back to school. Show students how to flip the pages to see the phases of the moon in action.

Astronomy

Constellation Books

Materials needed: 6" × 9" blue or black construction paper for each student
white crayon (or gold stick-on stars)
books about the stars that show diagrams of constellations

1. Have students work in pairs. Have each pair of students choose a different constellation.

2. On a piece of blue paper, each pair of students marks the constellation in pencil and then makes the stars using white crayon or gold stick-on stars. Label the constellation.

3. Compile the constellation diagrams into a booklet by stapling the papers together. Add construction paper covers and add this book to the science table and/or the classroom library.

4. Older students can research the constellations and write paragraphs telling where and when they appear in the night sky. Students can then use black markers to outline their constellations on empty oatmeal containers. After they poke holes along their outlines, they place small flashlights in the other ends. The constellations will be projected on the walls. Students can try to position their constellations in a classroom "planetarium."

Physical Sciences

The Study of How Things Work

PHYSICAL SCIENCES

Sound	44-47
Gravity	48-50
Light	51-53
Magnetism	54-56
Matter	57-60

Sound

Sound Is Vibration

The larynx, or voice box, is the main organ the body uses to produce sounds.

1. Have students place a hand on their throats to feel the vibrations as they speak.

2. Have students feel the difference in the vibrations when they speak high, low, loudly or softly.

The Sound Box

Materials needed: milk cartons
thin and thick rubber bands
scissors
crayons

1. Each student cuts a window in one side of a milk carton as shown.

2. He stretches a thin rubber band around the milk carton and plucks the rubber band. Students listen and watch what happens when they pluck their rubber bands. They do the same thing with their thick rubber bands.

3. Students can change the sound by placing a crayon under their rubber bands to make the rubber bands tighter. They can try this with several crayons.

4. Students will observe that their rubber bands vibrate when plucked. And the pitch varies according to the tautness and thickness of the rubber bands.

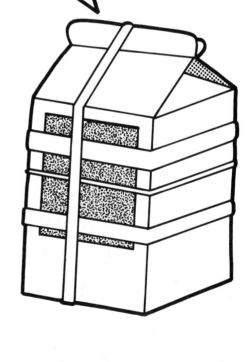

©Frank Schaffer Publications, Inc.   44   FS-8308 Instant Idea Book

Sound

Telephone Talk

Sound waves travel better through solids and liquids than through the air. String-and-cup telephones will illustrate this principle to your students. The string will carry sound better than air.

Materials needed: styrofoam cup for each student
heavy string

1. Have students work in pairs to make a string-and-cup telephone. They use the point of a pencil to make a small hole in the bottom of each cup. They thread each end of a string about twenty feet long through the holes in the cups. Then they make knots to secure their strings.

2. Have students stand apart holding their "telephones" so the string is taut. They take turns talking and listening to their partners. Then have students take one step toward each other so the string is no longer taut. Students should try to talk and listen on the "telephones." (When the string is not taut, they will not be able to hear each other as well.)

3. Let students have a four-way conversation by crisscrossing two telephones, making sure to loop one string around the other where they cross as shown. If strings are held taut, students can carry on a four-way conversation.

Criss-Cross String To Have Four-Way Conversation

Sound

Musical Bottles

Materials needed: five identical empty, glass, soda bottles
water

1. Do not put water in one bottle.
   Fill one bottle 1/4 full.
   Fill one bottle 1/2 full.
   Fill one bottle 3/4 full.
   Fill one bottle almost full.

2. Blow across the top of each bottle to make a whistling sound. The sounds will vary. Ask students which pitch is highest and lowest.

3. Have the students arrange the bottles in order from highest to lowest pitch. Tell students that the pitch is determined by the length of the air column in each bottle. The longer the column of air in the bottle, the lower the sound.

This experiment can also be done with identical water glasses. Instead of blowing, tap the rim of the glasses with a spoon.

The Incredible Shrinking Straw

Materials needed: one paper straw for each student
scissors

1. Have students flatten one end of their straws, and cut these ends to a point as shown.

2. They put the flat ends that they cut in their mouths. They blow through their straws and listen to the sound that is produced.

3. Next have students cut off approximately one inch from the round, unflattened ends of their straws. They blow again and listen to the new sound.

4. Students continue to cut off the ends of their straws in one-inch increments and blow through them. As their straws get shorter, the sound changes to a higher pitch.

Sound

Direct That Sound!

Sound waves can be directed. Each student will make and use a megaphone to control the direction of the sound waves.

Materials needed: 12" × 18" construction paper for each student

1. Students roll their papers to form cones.

2. Have students work in pairs and stand about ten feet apart. Students should repeat the same words or phrases, at the same volume, with and without their megaphones. Students should take turns talking and listening. They will be able to determine that when sound is directed through their megaphones, it is louder.

3. Tell students that the megaphone prevents the sound waves from dispersing rapidly. The megaphone helps the sound waves stay together longer which makes voices seem louder.

4. Now, show students that sound waves can be blocked. As a student is speaking through a megaphone, have him hold his hand or a book in front of the opening. The sound will be muffled because most of the sound waves will be stopped.

Gravity

Gravity

Gravity is a force that pulls things toward the earth. If there were no gravity, everything would float in the air.

Air Pressure Is Powerful

Demonstrate to students that air pressure can overcome the force of gravity.

Materials needed: a glass
water
a 3" x 5" card

1. Fill the glass with water.

2. Cover the top of the glass with the card and invert the glass as shown.

3. Remove your hand and the card will stay in place. The card and the water are not pulled down by gravity because the air pressure pushing against the card is stronger.

Controlling the Pull of Gravity

Show students how the shape of an object can sometimes slow the pull of gravity. For example, a parachute is pulled down by gravity at a slower rate because air pressure helps hold it up.

Materials needed: two 8 ½" x 11" pieces of paper.

1. Crumple one piece of paper into a ball.

2. Hold the paper ball in one hand and the flat sheet of paper in the other hand. Drop them both at the same time.

3. Observe the descent of the papers.

4. The flat paper floats down slowly because air pressure is helping to hold it up and the fact that the flat sheet's area is greater than the ball of paper.

Gravity

Falling Objects

An object increases its speed as it falls to the ground. The closer it gets to the ground, the faster it falls through the air. This demonstration will help your students understand this concept.

Materials needed: a small, heavy ball (golf ball, marble,...)
modeling clay

1. Make a flat clay pancake, approximately one inch thick, and place it on the floor.

2. Drop the ball from a height of three inches above the clay. Look at the indentation in the clay from the impact of the ball.

3. Now drop the ball from three feet into another clay pancake and examine the indentation.

4. Repeat this procedure from six feet.

5. Compare the three impressions. Determine which impression is deepest. Help students discover that the deepest impression is from the ball that fell the farthest because it hit the clay at a faster rate of speed. You can relate this to meteors which fall faster and cause deeper depressions in the earth.

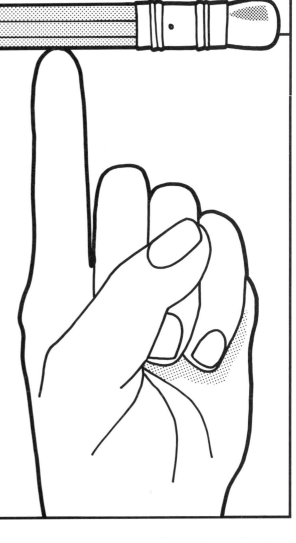

Finding the Center of Gravity

The center of gravity in an object is the point at which the most weight is concentrated.

Materials needed: pencil
ruler or yardstick for each student

1. Have each student find the center of gravity of a pencil by balancing it on one finger as shown.

2. Then have students find the center of gravity of other objects such as a ruler.

3. Explain to students that the pencil stays balanced as long as its center of gravity is directly over their fingers.

4. Suggest to students that they use their centers of gravity to balance when walking on a curb or balance beam. As long as they keep their centers of gravity directly above the curb, they will not fall.

Gravity

The Pull of Gravity

Show students how the force of gravity can pull an object.

Materials needed: one empty, half-gallon milk carton
a shoe box (without a lid)
string

1. Cut the top off the milk carton and attach string as shown.

2. Attach the other end of the string to the shoe box. Place the box on a table and suspend the carton over the edge.

3. Place a few objects (pencils, crayons, small book) in the suspended milk carton. When the shoe box begins to move toward the edge of the table, the force of gravity is pulling it.

4. The speed at which the shoe box moves across the table top can be varied by the weight of the objects in the shoe box or the milk carton.

Light

Shadows

This activity will show students how a shadow is formed when light is stopped by an object that will not let light through. (A shadow is the absence of light.)

Mystery Shadows

Materials needed: a flashlight or overhead projector

1. Darken the room. Hold up an object, such as a pair of scissors, in front of the light source. Cast the shadow on the wall or chalkboard.

2. Have students guess what the object is by looking at the shadow.

3. Let students discover that the shadow is shaped like the outline of the object because light cannot pass through it.

4. Continue to do this activity with a variety of objects.

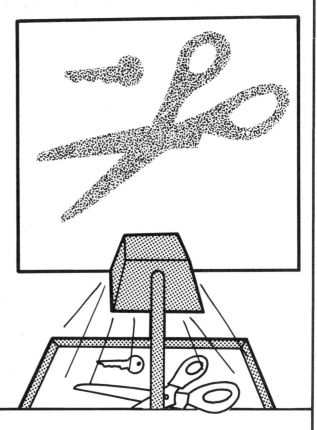

Shadow Portraits

Materials needed: filmstrip projector
black paper
tape

1. Have each student sit on a chair.

2. Tape a piece of black paper on the wall or chalkboard behind the student's head.

3. Shine the filmstrip projector on the student's profile so his shadow appears on the paper.

4. Trace around the shadow with a pencil and cut out the silhouette.

Light

Color and Light

Sunlight looks white, but it is actually made up of a spectrum of seven different colors. Rainbows appear when light is bent as it travels through raindrops. This causes white sunlight to be divided into the different colors of the spectrum. These colors always appear in the same order. Help students remember these colors and their positions with this activity.

The Colors of the Rainbow

Materials needed: 12″ × 18″ white drawing paper for each student
crayons

Have students use crayons to draw a rainbow. They should draw the bands of color in the order they appear:

    red
    orange
    yellow
    green
    blue
    indigo (deep blue, almost purple)
    violet

Make a Rainbow

Materials needed: a sunny day
a small mirror
a bowl of water
a sheet of white paper

1. Put a small mirror into a bowl of water so the sun shines on it.

2. Hold the white paper up so the mirror reflects on the paper. Show students the spectrum of colors that appears in rainbows.

Light

How Light Travels

Light travels through air in a straight line. Demonstrate to students that when light travels through air and then water, the light bends as it goes in and out of water.

Materials needed: glass of water
  pencil or straw

1. Put a pencil in a glass of water. Place the glass of water on a table or desk.

2. Have students look at the glass of water so it is level with their eyes. When they look into the glass, the pencil will look broken or bent.

3. Let students discover that the pencil looks broken because the light rays reflecting off the part of the pencil under water are bent.

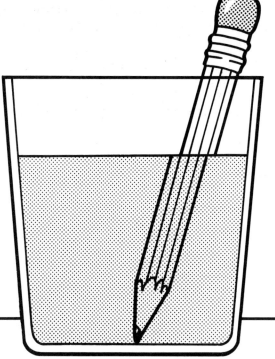

Light Can Be Reflected

Light reflected from a mirror bounces off at the same angle that the light strikes the mirror. (Some light "spreads out" and reaches your eyes.)

Materials needed: a ball
  a mirror
  a flashlight

1. Demonstrate to your class that when you throw a ball at an angle towards the ground, it bounces away at the same angle. Tell students that light can bounce off, or reflect, from a mirror in the same way.

2. Place the mirror on the floor. Darken the room.

3. Shine the flashlight on the mirror as shown. The light will be reflected from the mirror.

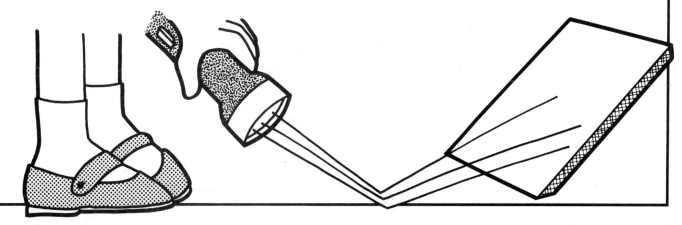

Magnetism

Pulling Power

Let students discover that magnets attract certain kinds of things.

Materials needed: assorted objects
two trays
a magnet

1. Mark one tray "Magnetic" and the other tray "Non-Magnetic."

2. During their free-time, students sort objects into the two trays.

3. Then let them use a magnet to determine if objects were sorted properly.

4. Have each student keep a list of which items he found "Magnetic" and "Non-Magnetic."

5. After all of your students have tested the items, compile a class list of "Magnetic" and "Non-Magnetic" objects. Encourage your students to explain why magnets attract only certain things.

Make a Magnet

Show students how they can use a magnet to magnetize another object for a short time.

Materials needed: a magnet
a nail or paper clip

1. Stroke the nail repeatedly in one direction with the magnet for one to two minutes.

2. Use the magnetized nail to pick up some pins.

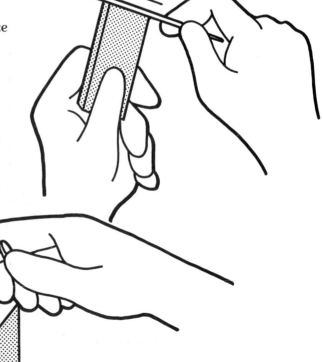

Magnetism

Magnets Are Powerful

Demonstrate to your students how the pulling force of a magnet can go through objects.

Materials needed: magnet
paper
drinking glass
foil
pin or paper clip

1. Place the pin on a tabletop. Cover it with a piece of paper.

2. Use the magnet's power to pick up the pin through the paper. When you pick up the pin, you will pick up the paper also.

3. Do this with two sheets of paper to test the power of the magnet.

4. Do this experiment using other materials to cover the pin.

5. Put the pin inside the glass. Place the magnet on the outside of the glass near the pin. The pin will be attracted to the magnet.

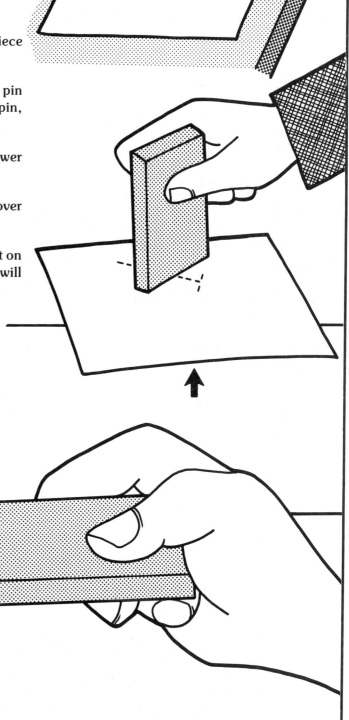

Magnetism

Magnets Attract and Repel Each Other

Show students how magnets have areas at each end where the magnetism is the strongest. These areas are called the poles. One end of a bar magnet suspended by a string will point north. This end of the magnet is called the north pole. The opposite end of the magnet is called the south pole.

Materials needed: two bar magnets
                          a string

1. Tie a string around the center of one bar magnet as shown. Have a student hold the string so the magnet is suspended in the air over a tabletop.

2. Have another student hold the other magnet near the suspended magnet. Have students observe what happens when different ends of the magnet are held near the suspended magnet.

3. Explain to students that if opposite poles of the magnets are near one another, they will attract. When like poles are near one another, the magnets move away from one another (repel).

Matter

Matter is all around you. Things that take up space are matter. You can touch matter. Everything that is matter has some weight. Matter can be in the form of a solid, liquid or gas.

Surface Tension

Liquid matter can be poured and always takes the shape of the container that holds it. This experiment will show students how the surface of liquids like water has a skin which is caused by **surface tension**.

Materials needed: glass
water
clean, dry paper clip
tweezers
liquid dishwashing detergent

1. Fill the glass to the brim with water. Use the tweezers to gently place a paper clip on the top of the water. The surface tension of the water has enough tension to support the paper clip.

2. Add a drop of liquid detergent to the water. Observe what happens.

3. Tell the students that the detergent reduces the surface tension causing the paper clip to sink.

Cup of Coins

Demonstrate how surface tension will stop the water from flowing out of a glass.

Materials needed: glass
water
pennies

1. Fill a glass completely with water.

2. Add coins one at a time to the glass of water. Lower each coin halfway into the water and then drop it as shown.

3. As you add coins, the water will bulge above the rim of the glass. Have students bend down to look across the top of the water to see the curve in the surface of the water.

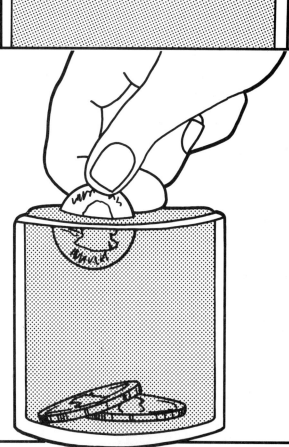

Matter

An Ocean of Air

Gas is matter that takes up space, has weight, and can be touched. Explain to students that air is all around them, even though it cannot be seen. Illustrate that air—a gas—takes up space and has weight by doing this experiment.

Push a glass down into a pan of water. The air inside the glass will keep the water out. Then tilt the glass and watch the air bubble out and some water go into the glass to take its place.

To demonstrate that air has weight, blow up two balloons. Tie a balloon on each end of a yardstick. Suspend the yardstick in the center from a string. Move the balloons so they are balanced and the stick is level. Pop one balloon allowing the air to escape. Now the stick will tilt toward the end with the balloon filled with air which is heavier.

TIE ON BALLOONS

Matter

Solid or Liquid?

Solids come in different shapes. Solid things cannot mix together like liquids. For example, your hand and a brick wall are solids; you cannot put your hand into the brick wall. Salt is a solid, but like a liquid, it can be poured. This experiment will show that salt is a solid.

Materials needed: salt
water
black construction paper
two small cups/glasses
magnifying glass/hand lens (optional)

1. Fill one cup with water. Let students observe how the water flows in and fills the cup.

2. Fill the other cup with salt. Have students observe how the salt pours in and fills the cup.

3. Let students look closely across the surface of both cups and observe how the water surface is smooth. When they look at the salt, they can see that it is made up of separate pieces.

4. Put a teaspoon of salt on the black paper. Now put a teaspoon of water in a different location on the black paper. Students will see that the salt is made up of many solid pieces.

Matter

Solids Can Hold Gases

Materials needed: a soft drink
   glass
   pan for water
   clump of dirt
   rock
   piece of brick
   scissors

1. Pour the soft drink into the glass. Let students observe the bubbles rising to the surface. These bubbles contain gas.

2. Explain to students that some solids are porous and contain gas. To illustrate this, fill the pan with water and drop in the piece of brick. Let students watch the bubbles of air that escape from the brick. Repeat the procedure with other classroom objects such as pencils, crayons,...

3. Help students conclude that objects that do not emit any bubbles were not holding any gas.

Matter Can Change Form

Demonstrate that matter can change from a solid, to a liquid and then to a gas.

Materials needed: a pie pan
   ice cube or block of ice

1. Place the ice cube in the pie pan. Now it is in the solid form.

2. Allow the ice to melt. When it melts, it has changed into a liquid form of matter.

3. Allow the water to evaporate. This changes the liquid matter into water vapor which is a gas.

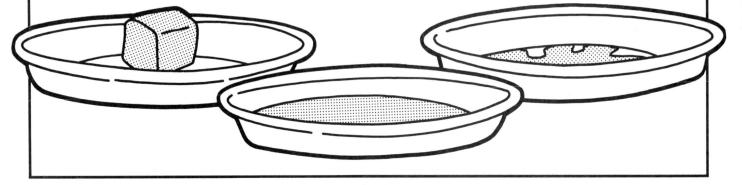

Name _____ Date _____

Science Experiment

1. Name of experiment: _____
   _____

2. I did an experiment to find out _____
   _____
   _____
   _____

3. This is what happened. _____
   _____
   _____
   _____
   _____

4. I learned _____
   _____
   _____
   _____
   _____

a reproducible page

Name _____ Date _____

Film Review

1. Film title: _____

2. It was about _____
   _____
   _____

3. I learned _____
   _____
   _____
   _____

Name _____ Date _____

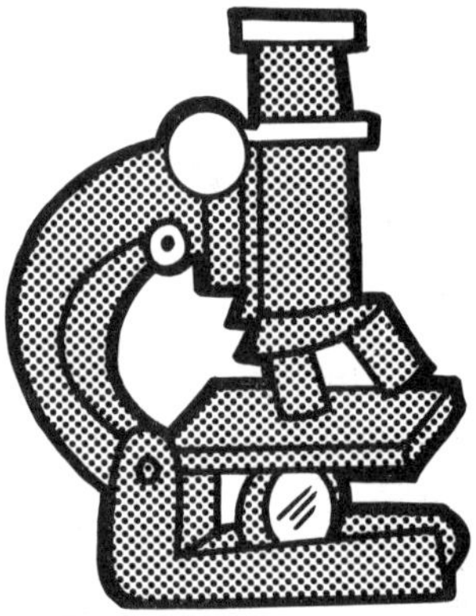

Science Reading Report

1. I read _____

2. It was about _____
   _____
   _____

3. I learned _____
   _____
   _____
   _____